# SCAVENGERS AND PARASITES

### in the Food Chain

**ALICE B. McGINTY**
**Photography by DWIGHT KUHN**

The Rosen Publishing Group's
**PowerKids Press™**
New York

*To my mother, Linda K. Blumenthal — Alice McGinty*

Published in 2002 by The Rosen Publishing Group, Inc.
29 East 21st Street, New York, NY 10010

First Edition

Book Design: Maria E. Melendez
Project Editor: Emily Raabe
Photo Credits: P. 11 © Doug Wechsler; all other photographs © Dwight Kuhn.

McGinty, Alice B.
Scavengers and parasites in the food chain / by Alice B. McGinty.
    p. cm. — (The Library of food chains and food webs)
Includes bibliographical references (p.      ).
  ISBN 0-8239-5755-1 (lib. bdg.)
1. Food chains (Ecology)—Juvenile literature. 2. Scavengers (Zoology)—Ecology—Juvenile literature. 3. Parasites—Ecology—Juvenile literature.
[1. Scavengers (Zoology) 2. Parasites. 3. Food chains (Ecology) 4. Ecology.] I. Title. II. Series.
  QH541 .M3854 2002
  577'.16—dc21

                                                                                    00–013028

Manufactured in the United States of America

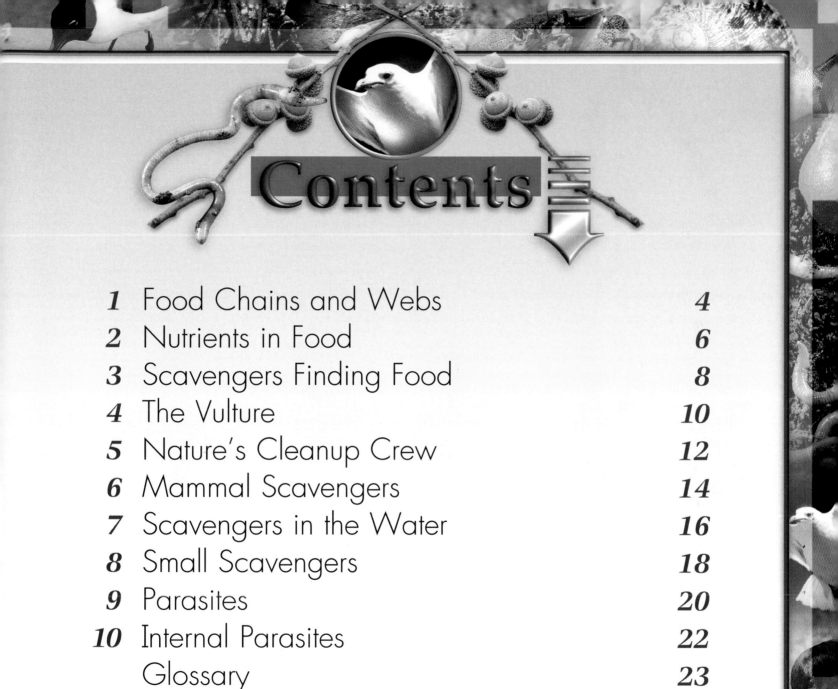

# Contents

# Food Chains and Webs

*A plant, a caterpillar, a robin, and a fox may form one food chain. Robins also eat worms and seeds. This makes robins part of several food chains. Foxes also eat mice and berries. This makes foxes part of many food chains, too. Most animals are part of many food chains. When animals are part of many food chains, they create a food web.*

When you are hungry, you may feel that you could eat almost anything. Scavengers are creatures that eat dead plants and animals. Parasites live off of other living creatures. You would not want to eat these foods, no matter how hungry you were! This book is about scavengers and parasites and the unusual foods that they eat.

Whenever an animal eats food, the animal and the food form a food chain. **Producers** are the first link in every food chain. Producers use the energy from sunlight to make their own food. **Consumers** in the food chain cannot make their own food. Consumers eat other living

4

things. Scavengers and parasites are both consumers. Every food chain ends with **decomposers**. Decomposers break down the bodies of dead plants and animals.

*This squirrel, a consumer, is eating fruit that comes from a plant. That plant is a producer.*

# Nutrients in Food

Each animal in the food chain gets **nutrients** and energy from the food it eats. Energy is passed along the food chain from one animal to another through nutrients. Some of the energy is used up as each animal grows and moves around. Animals must eat a lot of food to have enough energy to survive.

*The praying mantis is a dangerous carnivore, especially from the point of view of this elm leaf beetle that the mantis is about to eat!*

Animals that get their nutrients from plants are called **herbivores**. When a rabbit eats a plant, it gets nutrients from the plant. If a lion eats the rabbit, the lion gets nutrients from the rabbit's body.

6

Animals that get their nutrients from other animals are called **carnivores**. Some animals, such as foxes, eat both plants and animals. They are called **omnivores**. When animals die, scavengers may eat their bodies. Scavengers, such as vultures, get their nutrients from the bodies of dead plants and animals.

*This flatworm (shown eating a mosquito larva) is a scavenger.*

# Scavengers Finding Food

*Earthworms are both scavengers and decomposers in the food chain. Earthworms eat pieces of dead plants and animals that have been mixed in with the soil.*

Finding food is not always easy. In forests, grasslands, and oceans, many hungry animals compete for food. Scavengers have discovered a kind of food that is not eaten by most other animals. Instead of hunting and killing their food, scavengers eat animals and plants that have already died. Some scavengers eat dead animals. Other scavengers, such as earthworms, usually eat dead plant material. Scavengers that live in the ground also include millipedes, slugs, and snails.

Many herbivores, carnivores, omnivores, and decomposers are also scavengers. The arctic fox, for example, is an omnivore, but it will eat seal meat left over from a polar bear's meal. The wolf, a carnivore, hunts for live, sick, and dead animals. When an animal is hungry, it will eat almost any food it can find. The more kinds of food an animal will eat, the more likely it is to survive.

*Because they like to eat in the safety of their underground homes, earthworms pull their food underneath the soil. This helps fertilize the soil, making it rich for plants to grow.*

# The Vulture

Another name for a bird of prey is "raptor." Raptor comes from a Latin word meaning "to take and carry away." Most raptors, such as owls and eagles, are not scavengers. They use their strong talons to carry live prey back to their nests. All raptors have hooked beaks that help them tear meat.

Vultures are birds of **prey**. Birds of prey are birds that hunt and eat meat. Most birds of prey kill small animals with their sharp claws, called **talons**. However, vultures have weak talons that are not strong enough to kill prey. Vultures must find prey that already has been killed. This makes the vulture a scavenger.

The vulture's body has developed to make it a good scavenger. Vultures have long wings and sharp eyesight. They can soar in the sky for hours, looking for dead animals on the ground. Vultures also have bald heads and necks. To eat, vultures must stick their heads into the **carcass** of an animal. If a vulture had feathers on its head, germs from the carcass would cling to the feathers. The germs might make the vulture sick. The vulture's bald head and neck dry quickly so germs can't grow.

*The vulture's hooked beak is not strong enough to tear freshly killed meat. Sometimes vultures will poke at the body of a dead animal with their beaks to see if the meat has become soft enough to eat.*

# Nature's Cleanup Crew

Sometimes laughing gulls, such as this one, steal fish from pelicans. After a pelican catches a fish, it lifts its head from the water with the fish in its mouth. The laughing gull lands on the pelican's head, steals the fish, and flies away, making a screeching noise that sounds like laughter.

Scavengers clean up the earth by eating the bodies of dead plants and animals. Vultures soar above prairies, grasslands, and farmlands to find dead animals and eat them. Coyotes and wolves get rid of dead animals in forests. Many scavengers, such as crows and ravens, have learned to look for dead animals on roadways. Animals hit by cars on the highway are an easy source of food. Drivers are thankful that scavengers help keep the roads clean. Crows and ravens also look for dead fish that have washed up on seashores. Gulls and many other birds search the seashores for dead animals that have washed ashore. If it weren't for these birds, our beaches would be littered with smelly, dead fish. Scavengers play an important role in cleaning up the world.

*Crows eat many kinds of animals, both dead and alive. Crows kill and eat baby birds, fish, frogs, mice, insects, and small snakes.*

# Mammal Scavengers

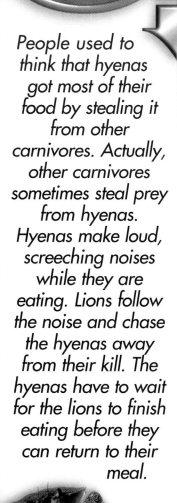

Mammals are animals that have hair and that feed milk to their young. Bears are mammals. Because they eat both plants and fish, bears are omnivores. Bears are also scavengers because they will eat dead fish and other animals when they find them. Many mammals are part-time scavengers.

Hyenas are well-known scavengers that live in Africa. They roam the grasslands and deserts looking for **carrion**. They eat the leftovers that lions, tigers, and leopards leave after killing and eating prey. Hyenas have the strongest jaws of any meat-eating animal. Hyenas use their strong jaws to eat the prey's

People used to think that hyenas got most of their food by stealing it from other carnivores. Actually, other carnivores sometimes steal prey from hyenas. Hyenas make loud, screeching noises while they are eating. Lions follow the noise and chase the hyenas away from their kill. The hyenas have to wait for the lions to finish eating before they can return to their meal.

bones and skin, which other carnivores usually leave behind. Hyenas also hunt and kill their own food. They hunt zebras, hares, and other animals. Hyenas tend to hunt in groups, called packs.

*Hyenas use their good sense of smell to find their prey, which is either dead or alive.*

# Scavengers in the Water

*Rock crabs live in tide pools. Tide pools are pools of water left when the level of the sea, or tide, is low. Rock crabs eat seaweed, worms, snails, clams, and pieces of dead plants and animals.*

Scavengers also help clean the oceans. Many ocean scavengers are **crustaceans**. Crustaceans have hard shells and bodies. Crabs are crustaceans. Some crabs live on beaches and eat dead plants and animals that have washed ashore. Other crabs live on the bottom of the sea and search for dead plants and animals that have sunk to the sea floor. Lobsters are another kind of crustacean scavenger that lives in the ocean. Lobsters use their strong claws to grab dead fish, live fish, smaller lobsters, and plants.

Crayfish, snails, worms, catfish, and clams are all pond, lake, and stream

16

scavengers. They eat the remains of dead plants and animals in the water. Crayfish live under stones or in deep burrows in the banks of streams.

*This crayfish is eating another crayfish that has died.*

# Small Scavengers

Slugs slide on the ground on a thin layer of mucus, which they make in their bodies. Slugs eat both living and dead plants, fungi, and sometimes, other dead slugs!

Many insects are scavengers. Some insect babies, called **larvae**, survive by eating dead plants and animals. When insects lay eggs, they work hard to be sure that their larvae will have food when they hatch. Burying beetles bury the bodies of dead animals. The female beetles lay their eggs in tunnels that lead to the buried bodies. When the eggs hatch, the larvae will eat the dead animals.

Dung beetles and their larvae eat the droppings of animals. Animal droppings, especially from herbivores, have pieces of food in them. The scarab beetle, a

18

kind of dung beetle, rolls animal droppings into a ball. Scarab beetles collect these balls in underground burrows, where they live. They lay their eggs inside the balls. When the eggs hatch, the larvae can eat the food in the animal droppings.

*This burying beetle has found a dead shrew. Now it must bury the shrew in the dirt to feed its larvae when they hatch.*

*The burying beetle has been hard at work, and the shrew is almost completely buried in the dirt. The beetle will continue to dig around the dead animal until it is totally buried.*

# Parasites

Parasites are living things that take food, and sometimes shelter, from other living things. Mosquitoes, for example, drink blood from people and other animals. The living things from which parasites get food or shelter are called hosts. The mosquito lives outside the body of its host. It is an **external parasite**. Ticks and fleas are external parasites, too. Different kinds of fleas prefer to live on different animals. Some fleas prefer dogs. Some fleas prefer cats. Others prefer birds, rats, or chipmunks. External parasites may bother their hosts, but they usually do

*Mosquitoes and fleas have sharp beaks to pierce the skin of their hosts. In their beaks are tubes. One tube pumps blood out of the host. The other tube pumps saliva into the host. The saliva keeps the host's blood thin so it doesn't clog up the tubes. The saliva is also what causes a mosquito or flea bite to itch.*

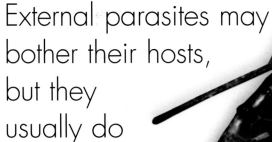

not kill them. A living host is more helpful to the parasite than a dead one! Sometimes, however, external parasites spread germs that kill their hosts. Mosquitoes can spread a dangerous disease called **malaria**. After a mosquito bites someone with malaria, it carries the germs to the next person it bites.

*Fleas can spread disease. Rat fleas were responsible for the spread of a disease called the plague, which killed many people in Europe in the 1500s.*

# Internal Parasites

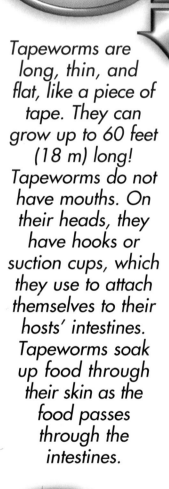

Tapeworms are long, thin, and flat, like a piece of tape. They can grow up to 60 feet (18 m) long! Tapeworms do not have mouths. On their heads, they have hooks or suction cups, which they use to attach themselves to their hosts' intestines. Tapeworms soak up food through their skin as the food passes through the intestines.

**Internal parasites** live inside the body of a host. Hookworms, pinworms, and tapeworms are internal parasites. They may live inside the intestines of pigs, birds, fish, or people. Internal parasites lay eggs inside their hosts. The eggs pass out of the hosts' bodies in their droppings. When the eggs hatch, the young parasites find new hosts.

Every member of the food chain needs food to survive. Scavengers and parasites, however, eat some foods that no other member of a food chain would eat. The next time you think that you are so hungry you could eat anything, remember these members of the food chain!

# Glossary

**carcass** (KAR-kus)  The body of a dead animal.

**carnivores** (KAR-nih-vorz)  Animals that eat other animals for food.

**carrion** (KEHR-ee-un)  Dead, rotting flesh of animals.

**consumers** (kon-SOO-mers)  Members of the food chain that eat other organisms.

**crustaceans** (krus-TAY-shunz)  Any group of animals that are invertebrates with hard shells, limbs, and antennae and live mostly in water.

**decomposers** (dee-kum-POH-zers)  Organisms that break down the bodies of dead plants and animals.

**external parasite** (ek-STUR-nel PAR-eh-syt)  An organism that lives or eats outside the body of the host organism.

**herbivores** (ER-bih-vorz)  Animals that eat plants.

**internal parasites** (in-TUHR-nel PAR-eh-syts)  Organisms that live and eat inside the body of the host organism.

**larvae** (LAHR-vee)  The form of a baby insect after it hatches from its egg and before it changes into an insect.

**malaria** (muh-LAR-ee-uh)  A serious disease common in very warm places, such as Africa.

**nutrients** (NOO-tree-ints)  Anything that living things need for energy or to grow.

**omnivores** (AHM-nih-vorz)  Animals that eat both plants and other animals.

**prey** (PRAY)  An animal that is hunted by another animal for food.

**producers** (pruh-DOO-serz)  Plants and algae that use sunlight to make their own food.

**talons** (TA-lunz)  Sharp, curved claws on a bird of prey.

23

# Index

# Web Sites

To learn more about scavengers and parasites in the food chain, check out these Web sites:

www.nationalgeographic.com/parasites
www.nhptv.org/natureworks/nwep11.htm